Make Music Count

Fractions Edition

Printed in the United States of America

First Printing, 2015

Make Music Count, LLC Atlanta, GA

www.MakeMusicCount.com

Contents

Abstract

In this workbook you will walk through piano lessons of great hip hop songs that will be derived from mathematical application. This workbook is your guide to applying mathematics to learn how to play the piano while also strengthening your understanding of basic fraction concepts. Every song in this journal will be learned by applying only mathematics.

Here is a picture of how the piano is normally taught where we are asked to view the notes as if we were looking at an actual piano.

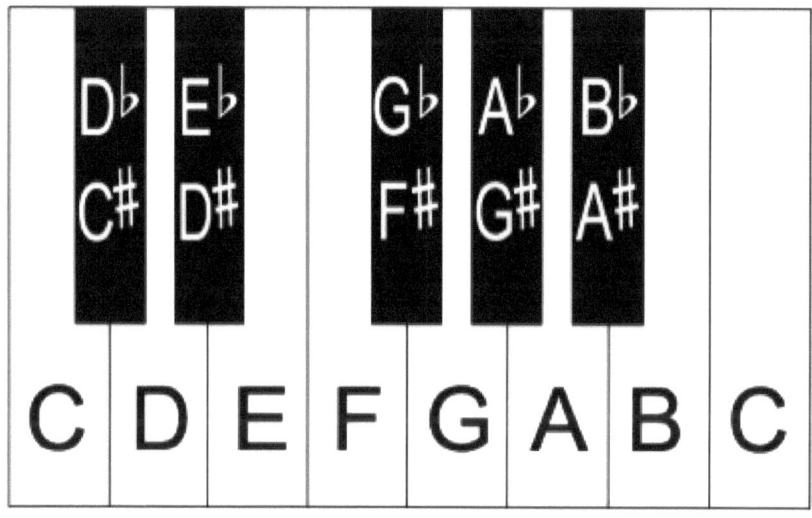

We will now look at the piano in a completely different manner; as a number line. Each musical note will now represent a point on a line. And the movement from one point on the line to another will signify the movement of playing one musical note to another. Here the concern is only the numerical distance between musical notes. By viewing the piano notes in this manner we will remove the music theory out of the music and be able to directly apply the notes on the number line to playing on the piano.

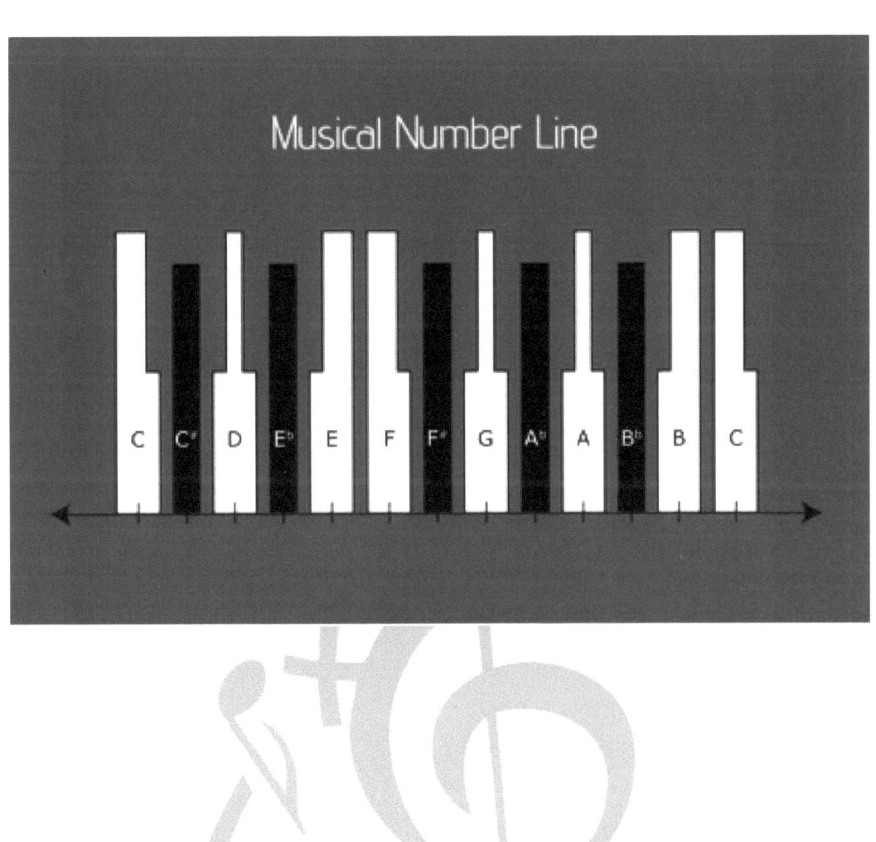

1 Half Steps

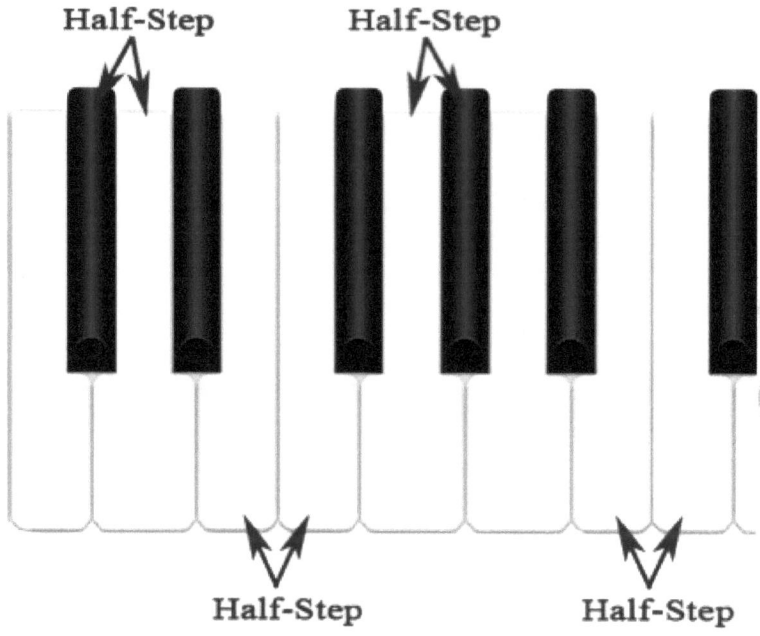

A half step in music is when you play one note on the piano and then move up or down to the very next note to play. For example if you played the note C, a half step up would be $C\sharp$, while a half step down from C would be the note B. On our musical number line we will represent a half step movement as 1/2. From this conclusion we see that the distance between each note on the Musical Number line is $\frac{1}{2}$.

Examples:

1. G $+ \frac{1}{2} = A\flat$
2. $B\flat - \frac{1}{2} =$ A

Half Step Practice

$C + \frac{1}{2} = $ _____

$A + \frac{1}{2} = $ _____

$C\sharp + \frac{1}{2} = $ _____

$E\flat + \frac{1}{2} = $ _____

$D\flat + \frac{1}{2} = $ _____

$G + \frac{1}{2} = $ _____

$E + \frac{1}{2} = $ _____

$F\sharp + \frac{1}{2} = $ _____

$B\flat + \frac{1}{2} = $ _____

$B + \frac{1}{2} = $ _____

$G\flat + \frac{1}{2} = $ _____

$F + \frac{1}{2} = $ _____

$A\sharp + \frac{1}{2} = $ _____

$D\sharp + \frac{1}{2} = $ _____

$F - \frac{1}{2} = $ _____

$B - \frac{1}{2} = $ _____

$A\flat - \frac{1}{2} = $ _____

$C - \frac{1}{2} = $ _____

$C\sharp - \frac{1}{2} = $ _____

$E\flat - \frac{1}{2} = $ _____

$G\flat - \frac{1}{2} = $ _____

$A - \frac{1}{2} = $ _____

$B\flat - \frac{1}{2} = $ _____

$E - \frac{1}{2} = $ _____

$D\sharp - \frac{1}{2} = $ _____

$A\sharp - \frac{1}{2} = $ _____

$F\sharp - \frac{1}{2} = $ _____

$G - \frac{1}{2} = $ _____

2 Whole Steps

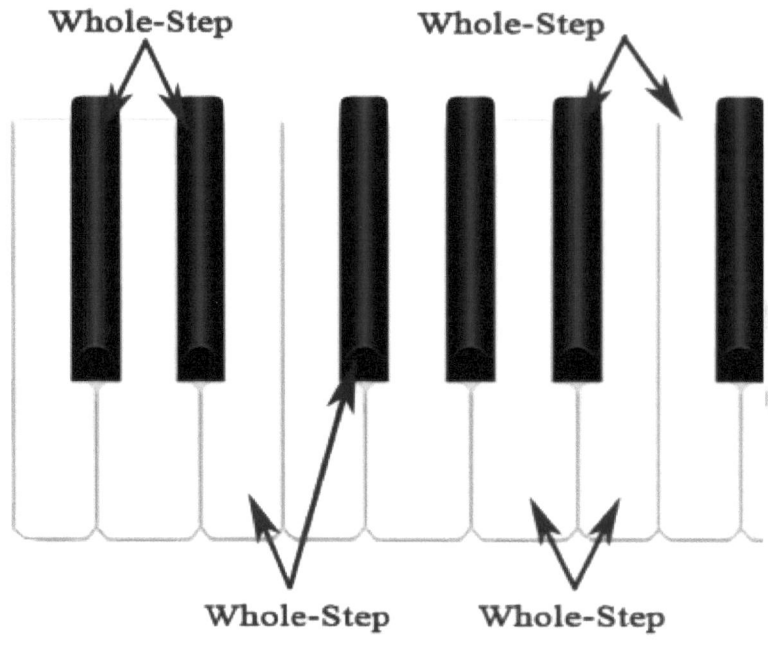

In music there are also whole steps. A whole step is when you play one note and you skip the very next note to play the following note. A whole step is also two half steps from the original starting note. An example of a whole step is if you played the note F and then moved and played the note G. Notice that in order to move from F to G the note F♯ must be skipped. We will define whole steps as 1.

Examples:

1. E + 1 = F♯
2. G + 2 = B

Whole Step Practice

C + 1 = _____

A + 1 = _____

C♯ + 1 = _____

E♭ + 1 = _____

D♭ + 1 = _____

G + 2 = _____

E + 2 = _____

F♯ + 2 = _____

B♭ + 2 = _____

B + 5 = _____

G♭ + 1 = _____

F + 2 = _____

A♯ + 4 = _____

D♯ + 3 = _____

F - 1 = _____

B - 1 = _____

A♭ - 1 = _____

C - 1 = _____

C♯ - 1 = _____

E♭ - 2 = _____

G♭ - 2 = _____

A - 2 = _____

B♭ - 2 = _____

E - 5 = _____

D♯ - 1 = _____

A♯ - 2 = _____

F♯ - 4 = _____

G - 3 = _____

9

3   Major Scales

Now that we can see how math is used to show the distance between musical notes, we can use the same application for deriving musical scales. Scales have a specific distance between each of the eight notes that make up that scale. With this knowledge we can derive any type of scale using math.

Let's take the major scales as an example. A major scale has eight notes, and when played it sounds "happy." Every note on the musical number line can serve as the beginning of a major scale, which means that there are twelve major scales to learn as an aspiring pianist. But here is the good news, each major scale uses the same sequence of math distances regardless of what note it starts on.

Here is the sequence to derive any major scale:

Major Scale = $1, 1, \frac{1}{2}, 1, 1, 1, \frac{1}{2}$

The numbers in the sequence represent the distance between each note of the major scale. Let's apply this sequence to derive the C Major Scale.

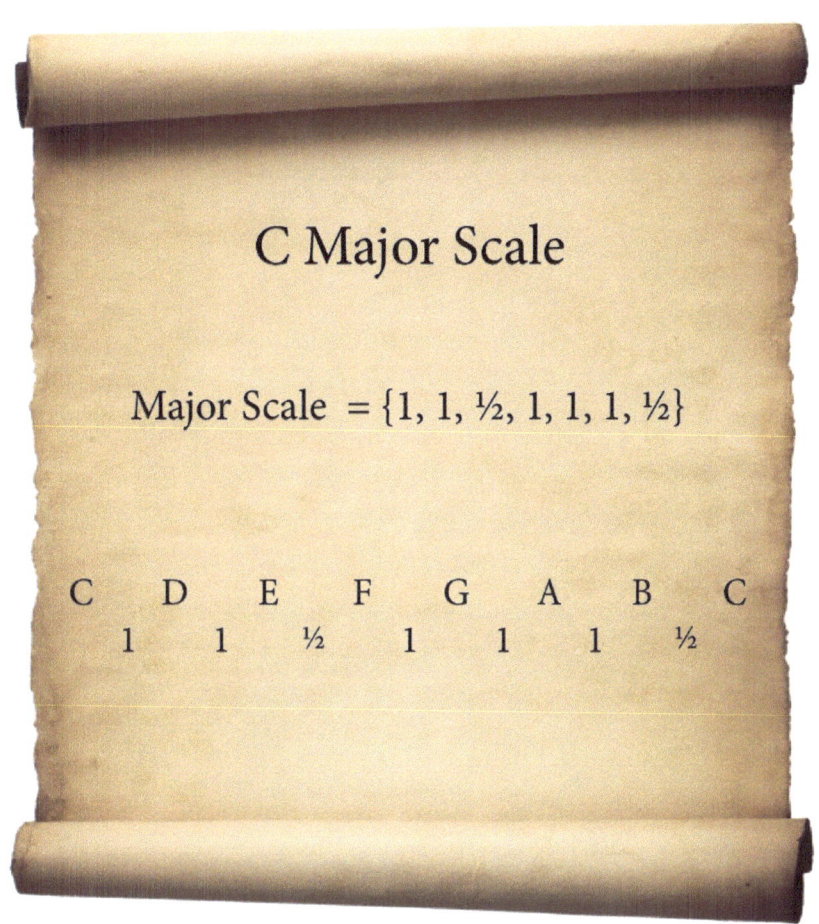

As you can see the number in the sequence represent the distance between each musical note. So now you are able to derive any major scale with this numerical sequence. All you have to do is pick a musical note to start with and use the sequence to determine the remaining seven musical notes in the major scale.

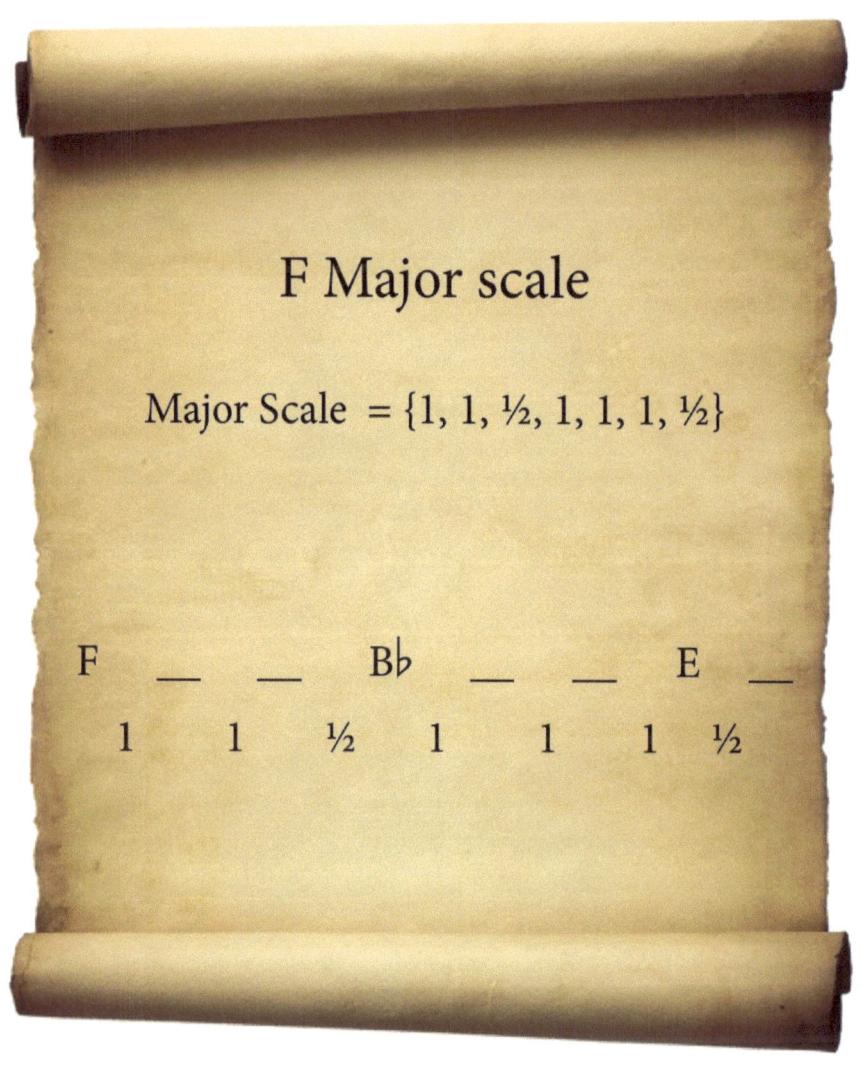

B♭ Major Scale

Major Scale = {1, 1, ½, 1, 1, 1, ½}

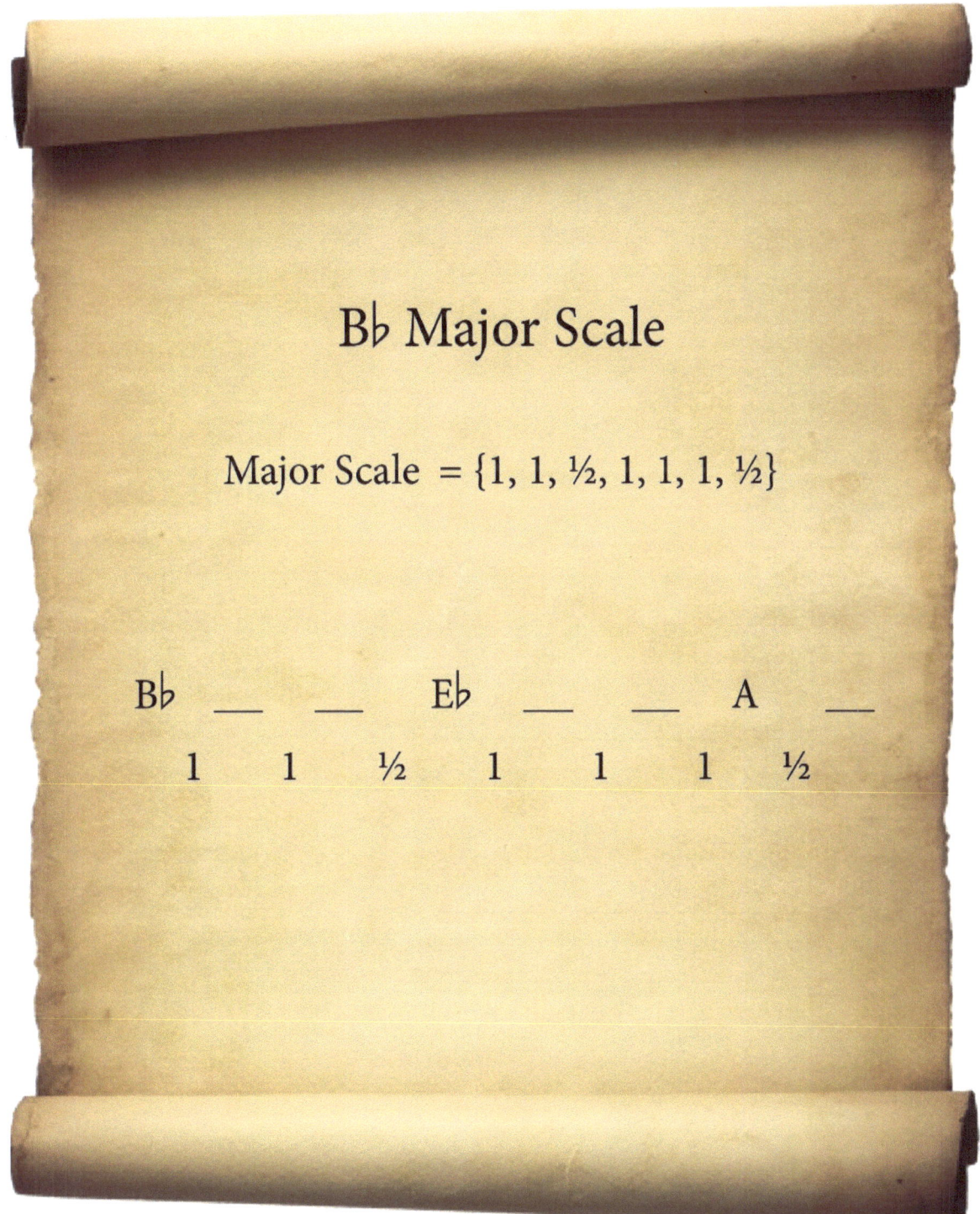

B♭ __ __ E♭ __ __ A __

1   1   ½   1   1   1   ½

E♭ Major Scale

Major Scale = {1, 1, ½, 1, 1, 1, ½}

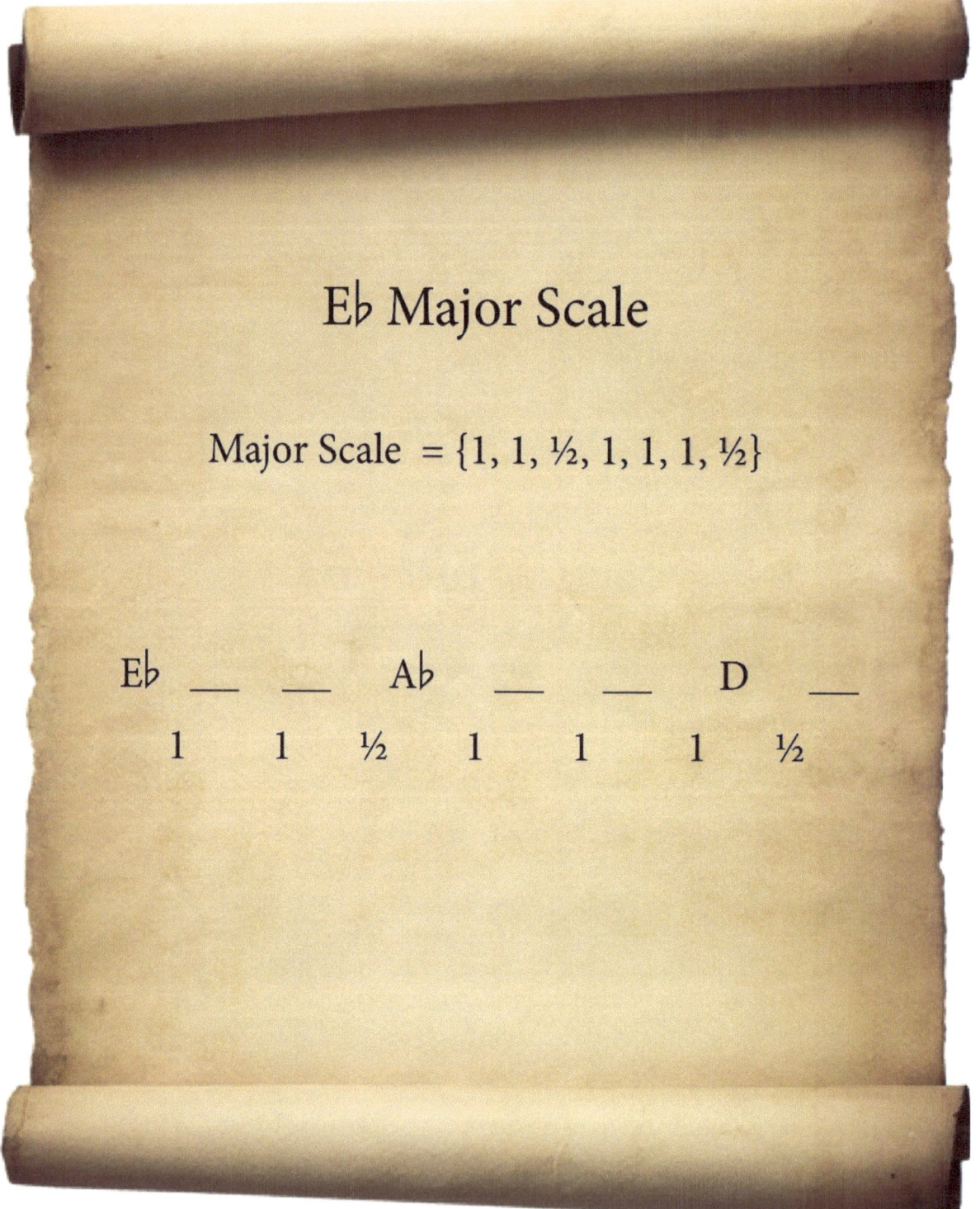

E♭ __ __ A♭ __ __ D __

1   1   ½   1   1   1   ½

A♭ Major Scale

Major Scale = {1, 1, ½, 1, 1, 1, ½}

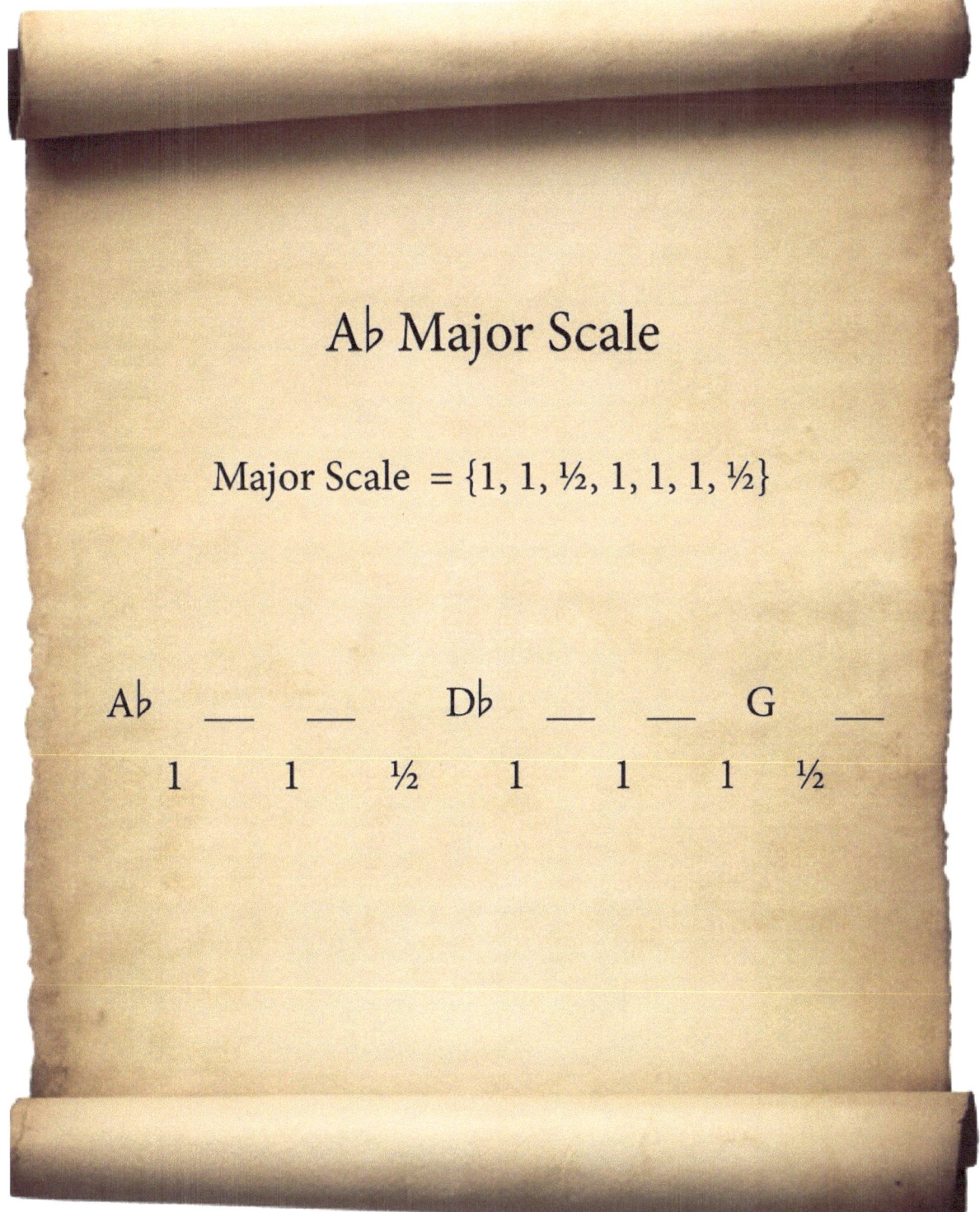

A♭ __ __ D♭ __ __ G __

1 1 ½ 1 1 1 ½

D♭ Major Scale

Major Scale $= \{1, 1, \frac{1}{2}, 1, 1, 1, \frac{1}{2}\}$

D♭ __ __ G♭ __ __ C __

  1   1   ½   1   1   1   ½

F# Major Scale

Major Scale = {1, 1, ½, 1, 1, 1, ½}

F# __ __ B __ __ E# __

1   1   ½   1   1   1   ½

B Major Scale

$$\text{Major Scale} = \{1, 1, \tfrac{1}{2}, 1, 1, 1, \tfrac{1}{2}\}$$

B __ __ E __ __ A# __

1　1　½　1　1　1　½

E Major Scale

Major Scale $= \{1, 1, \frac{1}{2}, 1, 1, 1, \frac{1}{2}\}$

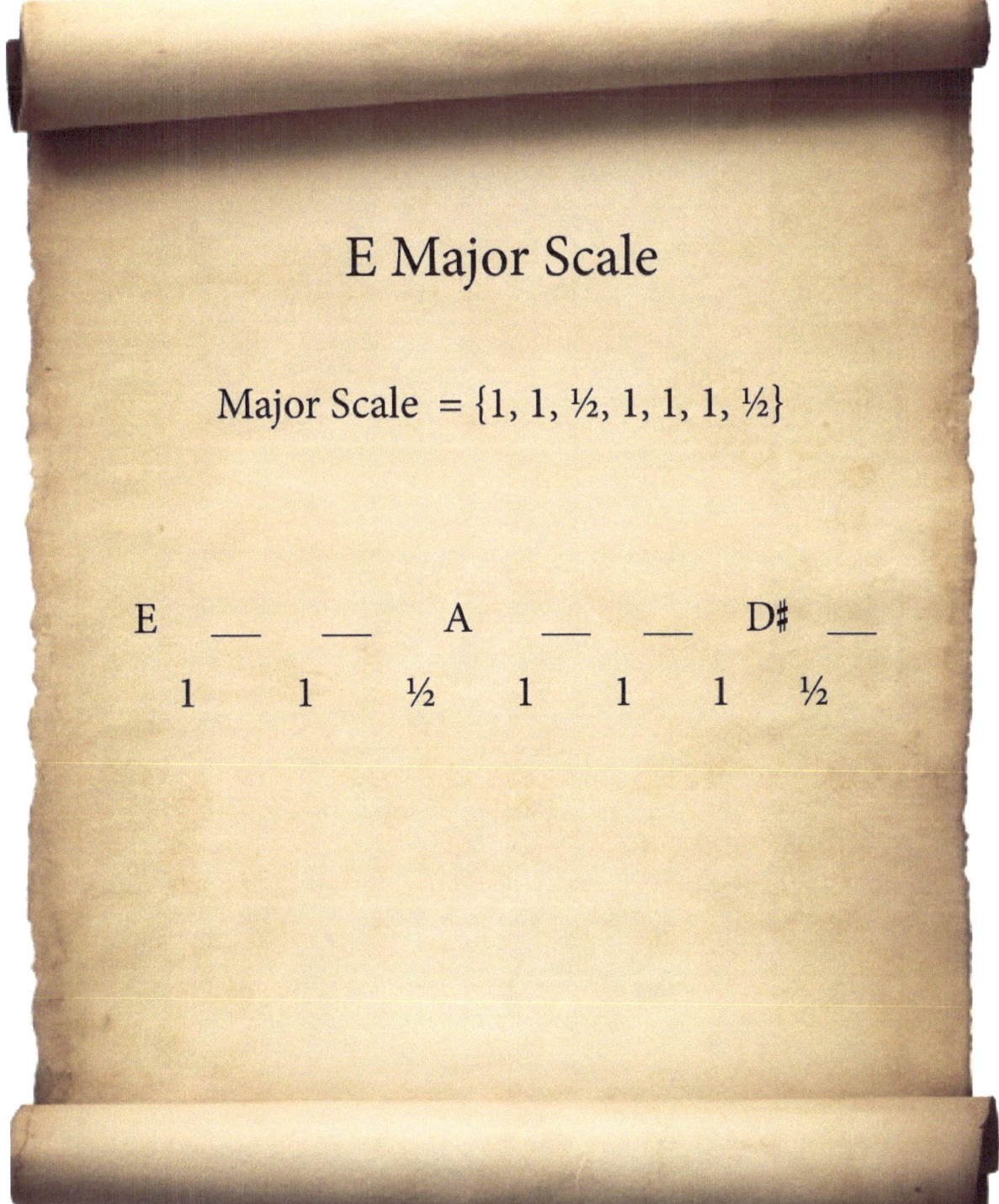

E __ __ A __ __ D# __

1　　1　　½　　1　　1　　1　　½

A Major Scale

Major Scale = {1, 1, ½, 1, 1, 1, ½}

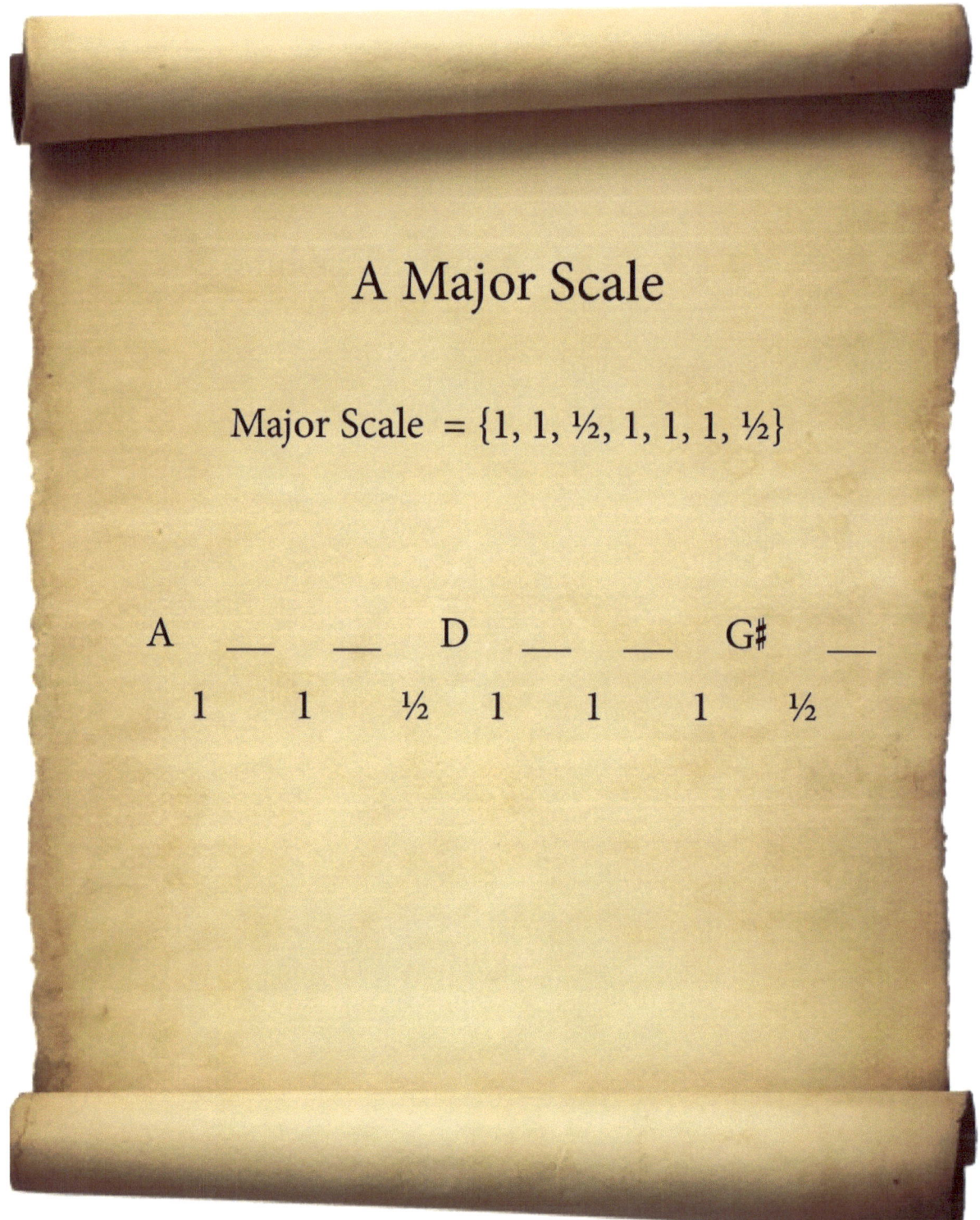

A __ __ D __ __ G♯ __

1   1   ½   1   1   1   ½

D Major Scale

Major Scale = {1, 1, ½, 1, 1, 1, ½}

D __ __ G __ __ C♯ __

1   1   ½   1   1   1   ½

G Major Scale

Major Scale $= \{1, 1, \frac{1}{2}, 1, 1, 1, \frac{1}{2}\}$

G __ __ C __ __ F# __

1   1   ½   1   1   1   ½

A benefit to learning scales through mathematical application instead of reading off a sheet of manuscript paper is that using the musical number line will force you to commit these scales to memory, which is a vital step for playing by ear. You will need to have your musical scales ready for you to match what you hear in music. The only way this is accomplished is by committing your scales to memory. As you begin to learn your scales you will recognize how the major scale sounds and you will be able to pick out the correct notes of other major scales without the help of the numerical sequence. Finding the correct notes in the scale is a great method for memorizing. The act of "picking out the correct notes" will force repetition, correction, and practice, which are key ingredients to memorizing scales. This method is also the beginning of playing by ear. You are finding the correct individual notes in scales by listening to what sounds right. Your ear will serve as the verification of what you derive through mathematics.

4 Adding and Subtracting Fractions

Now that you understand the mathematics behind deriving your scales, you will use the same method to derive the melodies of songs. You will be able to solve for the correct musical notes in order to play the following songs on the piano. The songs i this section will focus on adding and subtracting fractions in order to derive the melodies of the songs.

Imagine Dragons - Thunder

C   _____   _____

$-1\frac{1}{2}$ -2

Migos - Bad and Boujee

B♭   _____   _____

$-2\frac{1}{2}$ $\frac{1}{2}$

Miguel - Sky Walker

B   _____   _____

1                $-1\frac{1}{2}$

Chris Brown - Tempo

G $\qquad$ $\qquad$

-2 $\qquad$ $-\frac{1}{2}$

Imagine Dragons - Believer

B♭   _____   _____

    -2 $-\frac{1}{2}$

Rae Sremmurd - No Flex Zone

A♭ _____   _____   _____

6   -2 $-1\frac{1}{2}$

Coldplay -Viva La Vida

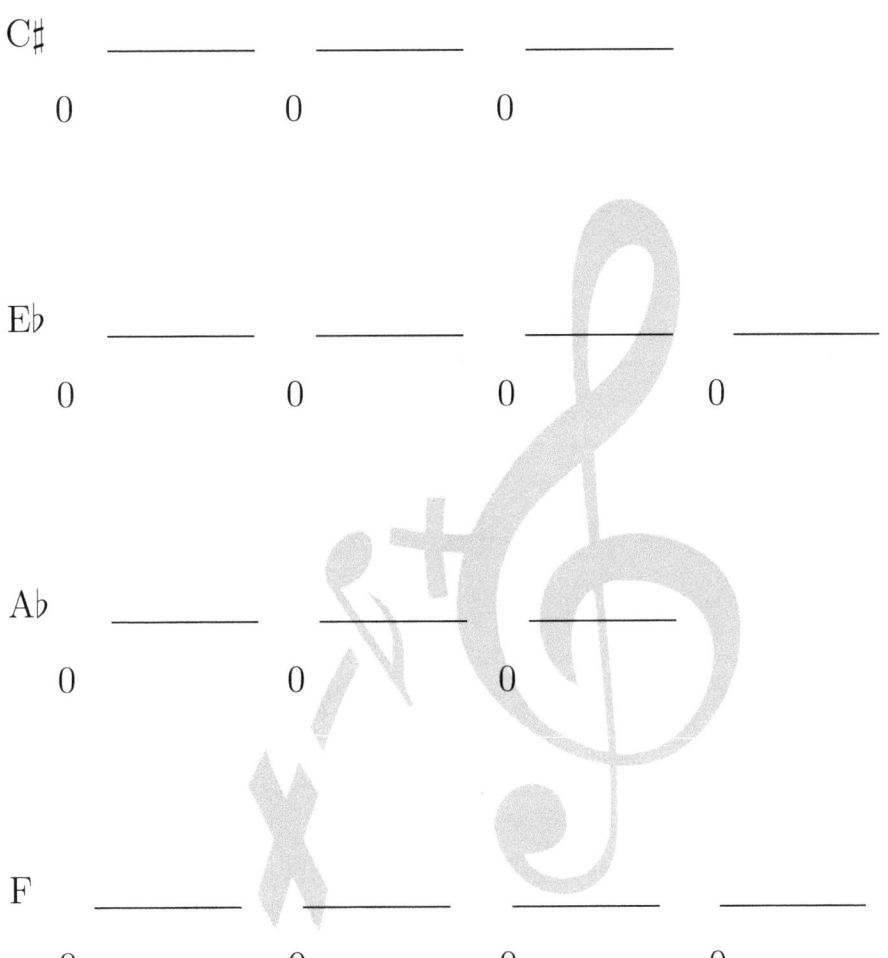

C♯ —————— —————— ——————

   0   0   0

E♭ —————— —————— —————— ——————

   0   0   0   0

A♭ —————— ——————

   0   0   0

F —————— —————— ——————

   0   0   0   0

Rich Homie Quan - Flex

F♯   _____   _____   _____

1                    $\frac{1}{2}$ -4

Rich The Kid - New Freezer

E _____   _____   _____   _____

$1\frac{1}{2}$ -7 $3\frac{1}{2}$ $2\frac{1}{2}$

J Balvin - Mi Gente

E _____ _____ _____ _____

$1\frac{1}{2}$ 2 $-2\frac{1}{2}$ 0

Ed Sheeran - Shape of You

Right Hand

C♯ _____ _____ _____ _____ _____

$1\frac{1}{2}$ $-1\frac{1}{2}$ 0 $1\frac{1}{2}$ $-1\frac{1}{2}$

C♯ _____ _____ _____ _____ _____

$1\frac{1}{2}$ $-1\frac{1}{2}$ 1 -1 -1

Left Hand

C♯ _____ _____ _____ _____ _____

0 0 $2\frac{1}{2}$ 0 0

A _____ _____ _____ _____ _____

0 0 1 0 0

Kendrick Lamar - Humble

E♭ _____ _____ _____ _____ _____

0        $\frac{1}{2}$ $-\frac{1}{2}$ $-3\frac{1}{2}$ 0

A♭ _____

4

Chris Brown - Party

C _____ _____ _____ _____ _____

0　　　2　　　0　　　$-2\frac{1}{2}$　　　0

E♭ _____

0

BlocBoy JB ft Drake - Look Alive

Right Hand

E _____   _____   _____   _____

    $\frac{1}{2}$ $-1\frac{1}{2}$ 1   -1

Left Hand

A _____   _____   _____   _____

    0   0   0 $\frac{1}{2}$

SZA - Love Galore

F   _____   _____   _____   _____   _____

$5\frac{1}{2}$ $2\frac{1}{2}$ $-4\frac{1}{2}$ $5\frac{1}{2}$ $2\frac{1}{2}$

Travis Scott - Antidote

F   _____   _____   _____   _____   _____

    -2   -1   2 $1\frac{1}{2}$ 1

Trinidad James - All Gold Everything

A   _____   G   A

$\frac{1}{2}$ _____   _____

D   E♭   _____   _____

$\frac{1}{2}$ $-1\frac{1}{2}$ 1

Kesha - TikTok

Right Hand

D _____ _____ _____ _____ _____
    0   0   0   1   0

F _____ _____
    0   0

Left Hand

B♭ _____ _____ _____ _____ _____
    0   0   0   1   0

D _____ _____
    0   0

Ne-Yo - So Sick

F♯   _____   _____   _____   _____   _____

$-\frac{1}{2}$ -2 $-1\frac{1}{2}$ -2 0

A♭   _____   _____

1 -1

Kendrick Lamar - M.A.A.d city

Ab   _____   _____   _____   _____   _____

0   0 $\frac{1}{2}$ 0   0

C♯   _____

0

Beyonce ft Nicki Minaj - Flawless

C   _____   _____   _____

    0        $1\frac{1}{2}$ 0

D   _____   _____   _____

  $1\frac{1}{2}$ $1\frac{1}{2}$ $-\frac{1}{2}$

2 Chainz - I'm Different

C _____ E _____ A _____ E

$-\frac{1}{2}$ _____ $2\frac{1}{2}$ _____ -2 _____

Lady Gaga - Applause

Right Hand

G _____ _____ B♭ _____ _____

   1        $\frac{1}{2}$ 0 $-\frac{1}{2}$ -1

Left Hand

G _____ _____ E♭ _____ _____

   -1   -1   0   1   1

Yo Gotti - Rake It Up

C♯   _____   _____   _____   _____

     0   1 $\frac{1}{2}$ 2

Nicki Minaj ft. Beyonce - Feeling Myself

E _____   G _____   G _____   D _____

1          $\frac{1}{2}$ 1   -1   1 $2\frac{1}{2}$ 1

Rich The Kid - New Freezer

E _____   _____   _____   _____

$1\frac{1}{2}$ -7 $3\frac{1}{2}$ $2\frac{1}{2}$

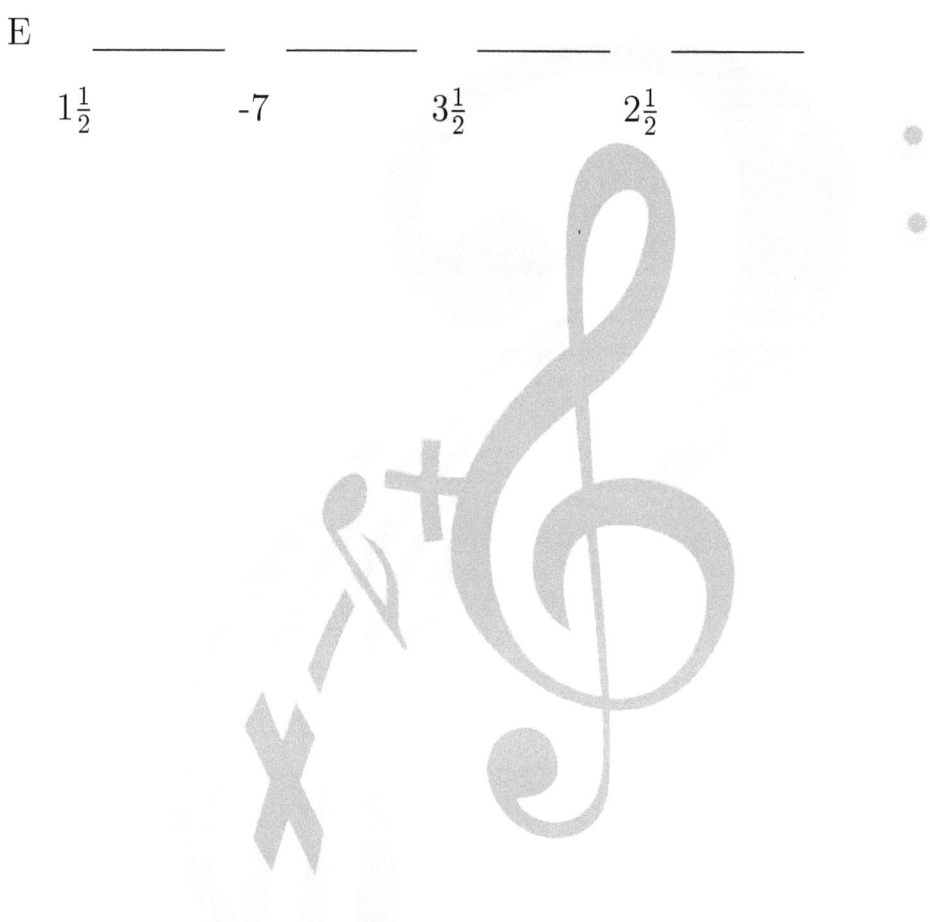

Maroon 5 ft SZA - What Lovers Do

E♭   ————   ————   ————   ————

    0   0   0   2

F   ————   ————   ————

    0   0   0

F   ————   ————   ————   ————

   $-3\frac{1}{2}$ 0   0   0

B♭   ————   ————   ————   ————

   $2\frac{1}{2}$ 0   0   0

Alicia Keys - Fallin'

Right Hand

G _____   E _____   B _____

   2 _____   0 _____   -2

Left Hand

E

Right Hand

A _____   F♯ _____   D _____

   $2\frac{1}{2}$ _____   0 _____ $-2\frac{1}{2}$

Left Hand

B

Migos - T-Shirt

B♭ $\frac{1}{2}$ _____ 0 _____ $-1\frac{1}{2}$ _____ $-2\frac{1}{2}$ _____

Charlie Puth - How Long

C♯   _____   _____   _____   _____   _____

     0   -1   0   -1   0

B   _____   _____

     0   0

Young Jeezy - R.I.P.

E _____   _____   F♯   _____

0              $1\frac{1}{2}$ _____ $-1\frac{1}{2}$

Macklemore - Thrift Shop

A♭ _____ _____ C♯ _____ _____

 6   -2½   -1   3½   3½

B _____

-1½

Journey - Don't Stop Believin

E   _____   _____   _____

    1   1 $1\frac{1}{2}$

C$\sharp$ _____   _____   _____

    1   -1   1

E   _____   _____   _____

  $-3\frac{1}{2}$ 3 $\frac{1}{2}$

N.E.R.D ft. Rihanna - Lemon

C   _____   _____   _____

    1   0 $\frac{1}{2}$

F   _____   _____   _____

    0   1 $\frac{1}{2}$

Fetty Wap ft. Remy Boyz - 679

G   ————   ————   F♯   ————   ————

0        $-\frac{1}{2}$ 0 $-3\frac{1}{2}$ 0

D   ————   ————

0   2

OG Maco - U Guessed It

B♭ _____ _____ _____
0   0 $2\frac{1}{2}$

B♭ _____ _____ _____
0   0 $\frac{1}{2}$

B♭ _____ _____ _____
3   -3 $2\frac{1}{2}$

B♭ _____ _____ _____
$2\frac{1}{2}$ -2 $-\frac{1}{2}$

B♭ _____ _____ _____
0          $\frac{1}{2}$ 0

B♭ _____ _____ _____ _____
0   0   3 $-\frac{1}{2}$

Meghan Trainor -All About That Bass

A   _____   _____   E   _____   _____

   1   1 $1\frac{1}{2}$ $-1\frac{1}{2}$ -2

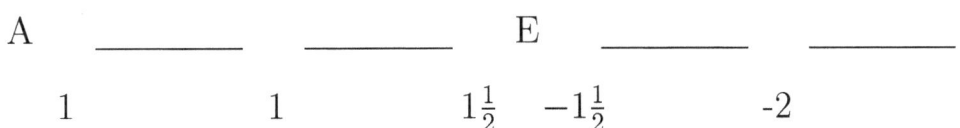

B   _____   _____   F♯   _____   _____

   1        $\frac{1}{2}$ 2   -2 $-1\frac{1}{2}$

E   _____   _____   B   _____   _____

   1   1 $1\frac{1}{2}$ $-1\frac{1}{2}$ -2

A   _____   _____   E   _____   _____

   1   1 $1\frac{1}{2}$ $-1\frac{1}{2}$ -2

Kendrick Lamar ft SZA - All the Stars

A♭   _____   _____   _____   _____   _____

0       $-1\frac{1}{2}$ $1\frac{1}{2}$ $-1\frac{1}{2}$ $3\frac{1}{2}$

G   _____   _____   _____

$2\frac{1}{2}$ -1 $-1\frac{1}{2}$

Rae Sremmurd - Throw Sum Mo'

A♭ _____ _____   E♭ _____ _____

-1   -1 $-\frac{1}{2}$ -1   1

E _____

1

A _____ _____   E♭ _____ _____

$-1\frac{1}{2}$ -1 $-\frac{1}{2}$ $-3\frac{1}{2}$ $2\frac{1}{2}$

E♭ _____

$\frac{1}{2}$

Migos ft. Drake - Walk it Talk it

D   _____   _____   _____   _____   _____

4 $-\frac{1}{2}$ 0 0 -2

Ariana Grande - One Last Time

A♭   _____   _____   _____

     1   1   -2

C♯   _____   _____   _____

    $-\frac{1}{2}$ $-3\frac{1}{2}$ 1

F   _____   _____   _____   _____

   -1   1 $1\frac{1}{2}$ $-\frac{1}{2}$

Zedd ft Maren Morris - The Middle

C   _____   _____   _____

    0              $-2\frac{1}{2}$ 0

D   _____   _____   _____

    0   -1   0

E   _____   _____   _____

    0   -1   0

5 Multiplying and Dividing Fractions

In this section we will now move towards the application of multiplying and dividing fractions. It is very important that you simply every fraction into its simplest form before you attempt to add that fraction to a musical note.

Pre Test

Please reduce all fractions to their lowest form

1. $\frac{1}{3}$ x $\frac{4}{5}$ =

2. $\frac{1}{2}$ ÷ $\frac{1}{4}$ =

3. $\frac{4}{4}$ x $\frac{3}{3}$ =

4. $\frac{1}{2}$ ÷ $\frac{1}{8}$ =

5. $\frac{1}{3}$ x $\frac{4}{5}$ =

6. $\frac{4}{5}$ ÷ $\frac{1}{3}$ =

7. $\frac{3}{4}$ x $\frac{2}{1}$ =

8. $\frac{1}{2}$ ÷ $\frac{1}{3}$ =

9. $\frac{7}{8}$ x $\frac{2}{4}$ =

10. $\frac{1}{2}$ ÷ $\frac{1}{16}$ =

Justin Timberlake - Mirrors

Right Hand

D + ($\frac{1}{2}$ x $\frac{2}{2}$) =   C$\sharp$ + ($\frac{1}{2}$ x $\frac{2}{2}$) =   B + ($\frac{1}{2}$ x $\frac{2}{2}$) =

B - ($\frac{1}{2}$ x $\frac{2}{2}$) =   C - ($\frac{2}{2}$ x $\frac{3}{3}$) =   F + ($\frac{3}{4}$ x $\frac{2}{1}$) =

A - ($\frac{1}{1}$ x $\frac{1}{1}$) =   E$\flat$ + ($\frac{2}{2}$ x $\frac{3}{3}$) =   B + ($\frac{3}{3}$ x $\frac{2}{1}$) =

Left Hand

E$\flat$ B$\flat$ A$\flat$

Rihanna - Diamonds

Right Hand

D♭ + ($\frac{1}{2}$ x $\frac{2}{2}$) =   F + ($\frac{1}{2}$ x $\frac{2}{2}$) =   E♭ + ($\frac{1}{2}$ x $\frac{2}{2}$) =

C - ($\frac{1}{2}$ x $\frac{2}{2}$) =   E - ($\frac{2}{2}$ x $\frac{3}{3}$) =   A♯ + ($\frac{3}{4}$ x $\frac{2}{1}$) =

A - ($\frac{1}{1}$ x $\frac{1}{1}$) =   G + ($\frac{2}{2}$ x $\frac{3}{3}$) =   F + ($\frac{3}{3}$ x $\frac{2}{1}$) =

Left Hand

G   B   A

Rich Gang - Lifestyle

Right Hand

A - ($\frac{1}{2}$ x $\frac{2}{2}$) =   E + ($\frac{1}{2}$ x $\frac{1}{1}$) =

E + ($\frac{1}{2}$ x $\frac{3}{3}$) =   B + ($\frac{1}{1}$ x $\frac{1}{1}$) =

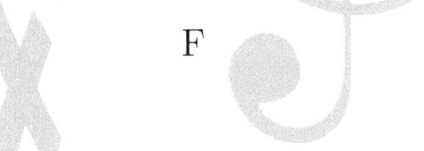

A + ($\frac{2}{1}$ x $\frac{3}{3}$) =   C - ($\frac{1}{1}$ x $\frac{2}{2}$) =

Left Hand

B♭   F   F♯

Rae Sremmurd - No Type

Right Hand

G - $\left(\frac{1}{2} \times \frac{2}{1}\right)$ =   Bb - $\left(\frac{1}{2} \times \frac{1}{1}\right)$ =

C♯ + $\left(\frac{1}{2} \times \frac{3}{3}\right)$ =   B + $\left(\frac{3}{1} \times \frac{1}{1}\right)$ =

D - $\left(\frac{2}{1} \times \frac{3}{3}\right)$ =   E - $\left(\frac{1}{1} \times \frac{2}{2}\right)$ =

Left Hand

Bb   D

Beyonce - 7/11

Right Hand

C + ($\frac{1}{2}$ ÷ $\frac{1}{4}$) =

A♭ - ($\frac{1}{4}$ ÷ $\frac{1}{8}$) =

A♭ - ($\frac{1}{4}$ ÷ $\frac{1}{16}$) =

G♯ + ($\frac{2}{2}$ ÷ $\frac{1}{2}$) =

F + ($\frac{1}{4}$ ÷ $\frac{1}{8}$) =

C♯ - ($\frac{2}{2}$ ÷ $\frac{1}{2}$) =

Left Hand

A

F

Drake - Hotline Bling

Right Hand

F♯ + ($\frac{1}{2}$ ÷ $\frac{1}{4}$) =   C♯ - ($\frac{1}{4}$ ÷ $\frac{1}{8}$) =

D♭ - ($\frac{1}{4}$ ÷ $\frac{1}{16}$) =   C + ($\frac{2}{2}$ ÷ $\frac{1}{2}$) =

B♭ + ($\frac{1}{4}$ ÷ $\frac{1}{8}$) =   E - ($\frac{2}{2}$ ÷ $\frac{1}{2}$) =

Left Hand

B♭   A

Miley Cyrus - Party in the USA

Right Hand

A + ($\frac{1}{2}$ x $\frac{2}{2}$) =   G + ($\frac{1}{2}$ x $\frac{2}{2}$) =   A + ($\frac{1}{2}$ x $\frac{2}{2}$) =

G - ($\frac{1}{2}$ x $\frac{2}{2}$) =   G - ($\frac{2}{2}$ x $\frac{3}{3}$) =   E♭ + ($\frac{3}{4}$ x $\frac{2}{1}$) =

Left Hand

F♯   B♭   E♭

Sam Smith - Stay With Me

Right Hand

$A\flat + (\frac{1}{2} \times \frac{2}{2}) =$ 　　　　$E + (\frac{1}{2} \times \frac{2}{2}) =$ 　　　　$E\flat + (\frac{1}{2} \times \frac{2}{2}) =$

$F - (\frac{1}{2} \times \frac{2}{2}) =$ 　　　　$D - (\frac{2}{2} \times \frac{3}{3}) =$ 　　　　$A + (\frac{3}{4} \times \frac{2}{1}) =$

$D - (\frac{1}{1} \times \frac{1}{1}) =$ 　　　　$G + (\frac{2}{2} \times \frac{3}{3}) =$ 　　　　$E\flat + (\frac{3}{3} \times \frac{2}{1}) =$

Left Hand

A 　　　　　　　　　F 　　　　　　　　　C

Fety Wap - Trap Queen

Right Hand

$B + (\frac{1}{2} \times \frac{2}{2}) =$ $\qquad$ $A\sharp + (\frac{1}{2} \times \frac{2}{2}) =$ $\qquad$ $B\flat + (\frac{1}{2} \times \frac{2}{2}) =$

$A\sharp - (\frac{1}{2} \times \frac{2}{2}) =$ $\qquad$ $A - (\frac{2}{2} \times \frac{3}{3}) =$ $\qquad$ $E + (\frac{3}{4} \times \frac{2}{1}) =$

$F\sharp - (\frac{1}{1} \times \frac{1}{1}) =$ $\qquad$ $D + (\frac{2}{2} \times \frac{3}{3}) =$ $\qquad$ $C + (\frac{3}{3} \times \frac{2}{1}) =$

Left Hand

A $\qquad$ G $\qquad$ E

Lynyrd Skynyrd - Sweet Home Alabama

Right Hand

$G\sharp + (\frac{1}{2} \times \frac{2}{2}) =$ $F\sharp + (\frac{1}{2} \times \frac{2}{2}) =$ $G\flat + (\frac{1}{2} \times \frac{2}{2}) =$

$G - (\frac{1}{2} \times \frac{2}{2}) =$ $F\sharp - (\frac{2}{2} \times \frac{3}{3}) =$ $B + (\frac{3}{4} \times \frac{2}{1}) =$

$E - (\frac{1}{1} \times \frac{1}{1}) =$ $B\flat + (\frac{2}{2} \times \frac{3}{3}) =$ $G + (\frac{3}{3} \times \frac{2}{1}) =$

Left Hand

D   C   G

Taylor Swift - Shake it Off

Right Hand

E♭ + ($\frac{1}{2} \div \frac{1}{4}$) =   A - ($\frac{1}{2} \div \frac{1}{2}$) =   A♭ - ($\frac{1}{2}$ x $\frac{2}{2}$) =

F♯ - ($\frac{4}{4}$ x $\frac{2}{2}$) =   C + ($\frac{1}{2} \div \frac{1}{4}$) =   E - ($\frac{1}{2} \div \frac{1}{2}$) =

G♯ - ($\frac{1}{2} \div \frac{1}{8}$) =   A♭ + ($\frac{1}{2} \div \frac{1}{4}$) =   D - ($\frac{3}{4}$ x $\frac{2}{1}$) =

Left Hand

A   C   G

O.T. Genasis - Coco

B♭ + ($\frac{1}{2}$ ÷ $\frac{1}{4}$) =   F - ($\frac{1}{2}$ ÷ $\frac{1}{2}$) =   E - ($\frac{2}{2}$ x $\frac{2}{2}$) =

C - ($\frac{4}{4}$ x $\frac{2}{2}$) =   F♯ + ($\frac{1}{2}$ ÷ $\frac{1}{4}$) =   F + ($\frac{1}{2}$ ÷ $\frac{1}{4}$) =

B + ($\frac{1}{2}$ ÷ $\frac{1}{8}$) =   B - ($\frac{1}{2}$ ÷ $\frac{1}{4}$) =   E + ($\frac{3}{4}$ x $\frac{2}{1}$) =

Maroon 5 - Animals

Right Hand

B + ($\frac{1}{2}$ ÷ $\frac{1}{8}$) =   D + ($\frac{1}{2}$ ÷ $\frac{1}{4}$) =   C + ($\frac{4}{3}$ x $\frac{3}{2}$) =

E - ($\frac{4}{4}$ x $\frac{2}{2}$) =   F♯ - ($\frac{1}{2}$ ÷ $\frac{1}{4}$) =   A♭ + ($\frac{1}{2}$ ÷ $\frac{1}{4}$) =

G + ($\frac{1}{2}$ ÷ $\frac{1}{4}$) =   B - ($\frac{1}{3}$ ÷ $\frac{1}{3}$) =   A - ($\frac{2}{2}$ ÷ $\frac{2}{2}$) =

Left Hand

E   D   C

Ariana Grande - Break Free ft Zedd

Right Hand

$B + (\frac{1}{2} \div \frac{1}{8}) =$ $F\sharp + (\frac{1}{2} \div \frac{1}{4}) =$ $F + (\frac{4}{3} \times \frac{3}{2}) =$

$F - (\frac{4}{4} \times \frac{2}{2}) =$ $A - (\frac{1}{2} \div \frac{1}{4}) =$ $D\flat + (\frac{1}{2} \div \frac{1}{4}) =$

$F\sharp + (\frac{1}{2} \div \frac{1}{4}) =$ $E - (\frac{1}{3} \div \frac{1}{3}) =$ $D - (\frac{2}{2} \div \frac{2}{2}) =$

Left Hand

E$\flat$ G   F

Usher - I Don't Mind

Right Hand

F♯ + ($\frac{1}{2}$ x $\frac{2}{1}$) =   B - ($\frac{1}{2}$ ÷ $\frac{1}{8}$) =   D - ($\frac{1}{2}$ x $\frac{6}{1}$) =   B + ($\frac{1}{2}$ ÷ $\frac{1}{4}$) =

D + ($\frac{1}{2}$ x $\frac{2}{1}$) =   G - ($\frac{1}{2}$ ÷ $\frac{1}{8}$) =   B♭ - ($\frac{1}{2}$ x $\frac{6}{1}$) =   G + ($\frac{1}{2}$ ÷ $\frac{1}{4}$) =

A + ($\frac{1}{2}$ x $\frac{2}{1}$) =   D - ($\frac{1}{2}$ ÷ $\frac{1}{8}$) =   F - ($\frac{1}{2}$ x $\frac{6}{1}$) =   D + ($\frac{1}{2}$ ÷ $\frac{1}{4}$) =

Left Hand

E   C♯   A   A♭

Taylor Swift - Bad Blood

Right Hand

F + ($\frac{1}{2}$ x $\frac{2}{1}$) =   E$\flat$ - ($\frac{1}{2}$ ÷ $\frac{1}{8}$) =   C - ($\frac{1}{2}$ x $\frac{6}{1}$) =   B + ($\frac{1}{2}$ ÷ $\frac{1}{4}$) =

D + ($\frac{1}{2}$ x $\frac{2}{1}$) =   G - ($\frac{1}{2}$ ÷ $\frac{1}{8}$) =   A$\flat$ - ($\frac{1}{2}$ x $\frac{6}{1}$) =   G + ($\frac{1}{2}$ ÷ $\frac{1}{4}$) =

A$\sharp$ + ($\frac{1}{2}$ x $\frac{2}{1}$) =   B$\flat$ - ($\frac{1}{2}$ ÷ $\frac{1}{8}$) =   D$\sharp$ - ($\frac{1}{2}$ x $\frac{6}{1}$) =   B$\flat$ + ($\frac{1}{2}$ ÷ $\frac{1}{4}$) =

Left Hand

C   G   D   E

Ed Sheeran- Thinking Out Loud

Right Hand

$D + (\frac{4}{3} \times \frac{3}{2}) =$ $B\flat - (\frac{1}{2} \div \frac{1}{4}) =$ $C\sharp - (\frac{1}{2} \times \frac{6}{1}) =$ $E\flat + (\frac{1}{2} \div \frac{1}{4}) =$

$B\flat + (\frac{2}{2} \times \frac{2}{1}) =$ $F\sharp - (\frac{1}{4} \div \frac{1}{8}) =$ $A\flat - (\frac{1}{2} \times \frac{6}{1}) =$ $B\flat + (\frac{1}{2} \div \frac{1}{4}) =$

$F\sharp + (\frac{1}{2} \times \frac{3}{1}) =$ $F - (\frac{1}{2} \div \frac{1}{8}) =$ $E - (\frac{1}{2} \times \frac{5}{1}) =$ $G + (\frac{1}{2} \div \frac{1}{4}) =$

Left Hand

D   F♯   G   A

Major Lazer - Lean On

Right Hand

$B\flat + (\frac{4}{3} \times \frac{3}{2}) =$ $E - (\frac{1}{2} \div \frac{1}{4}) =$ $G\sharp - (\frac{1}{2} \times \frac{6}{1}) =$ $B\flat + (\frac{1}{2} \div \frac{1}{4}) =$

$F\sharp + (\frac{2}{2} \times \frac{2}{1}) =$ $C\sharp - (\frac{1}{4} \div \frac{1}{8}) =$ $E - (\frac{1}{2} \times \frac{6}{1}) =$ $G\flat + (\frac{1}{2} \div \frac{1}{4}) =$

$E + (\frac{1}{2} \times \frac{3}{1}) =$ $C\sharp - (\frac{1}{2} \div \frac{1}{8}) =$ $C - (\frac{1}{2} \times \frac{5}{1}) =$ $E\flat + (\frac{1}{2} \div \frac{1}{4}) =$

Left Hand

$E\flat$ F   G $B\flat$

Adele - Hello

Right Hand

$E + (\frac{4}{3} \times \frac{3}{2}) =$ $C - (\frac{1}{2} \div \frac{1}{4}) =$ $D - (\frac{1}{2} \times \frac{6}{1}) =$ $E\flat + (\frac{1}{2} \div \frac{1}{4}) =$

$E\flat + (\frac{1}{2} \times \frac{2}{1}) =$ $C\sharp - (\frac{1}{4} \div \frac{1}{16}) =$ $F\sharp - (\frac{1}{4} \times \frac{6}{1}) =$ $B + (\frac{1}{2} \div \frac{1}{4}) =$

$A + (\frac{1}{2} \times \frac{3}{1}) =$ $A - (\frac{1}{2} \div \frac{1}{8}) =$ $F - (\frac{1}{2} \times \frac{5}{1}) =$ $F\sharp + (\frac{1}{2} \div \frac{1}{4}) =$

Left Hand

F   C♯   A♭   E♭

Disclosure ft Sam Smith - Latch

Right Hand

$E + (\frac{3}{2} \times \frac{4}{3}) =$ $C - (\frac{1}{2} \div \frac{1}{4}) =$ $D - (\frac{1}{2} \times \frac{6}{1}) =$ $D\flat + (\frac{1}{2} \div \frac{1}{4}) =$

$D + (\frac{1}{2} \times \frac{3}{1}) =$ $B - (\frac{1}{2} \div \frac{1}{8}) =$ $G\sharp - (\frac{1}{2} \times \frac{5}{1}) =$ $A + (\frac{1}{2} \div \frac{1}{4}) =$

$B + (\frac{1}{2} \times \frac{2}{1}) =$ $A\flat - (\frac{1}{4} \div \frac{1}{16}) =$ $D\sharp - (\frac{1}{4} \times \frac{6}{1}) =$ $A\flat + (\frac{1}{2} \div \frac{1}{2}) =$

Left Hand

B♭   F   E♭   C

Post Test

Please reduce all fractions to their lowest form

1. $\frac{1}{3}$ x $\frac{4}{5}$ =

2. $\frac{1}{2}$ ÷ $\frac{1}{4}$ =

3. $\frac{4}{4}$ x $\frac{3}{3}$ =

4. $\frac{1}{2}$ ÷ $\frac{1}{8}$ =

5. $\frac{1}{3}$ x $\frac{4}{5}$ =

6. $\frac{4}{5}$ ÷ $\frac{1}{3}$ =

7. $\frac{3}{4}$ x $\frac{2}{1}$ =

8. $\frac{1}{2}$ ÷ $\frac{1}{3}$ =

9. $\frac{7}{8}$ x $\frac{2}{4}$ =

10. $\frac{1}{2}$ ÷ $\frac{1}{16}$ =